Clinton Thomas Dent

The Nature and Significance of Pain Considered in its Physiological Aspect

An introductory address delivered at St. George's Hospital, on 3rd

October, 1887

Clinton Thomas Dent

The Nature and Significance of Pain Considered in its Physiological Aspect
An introductory address delivered at St. George's Hospital, on 3rd October, 1887

ISBN/EAN: 9783337013271

Printed in Europe, USA, Canada, Australia, Japan

Cover: Foto ©berggeist007 / pixelio.de

More available books at **www.hansebooks.com**

THE NATURE

AND

SIGNIFICANCE OF PAIN

CONSIDERED IN ITS PHYSIOLOGICAL ASPECT.

AN INTRODUCTORY ADDRESS

DELIVERED AT ST. GEORGE'S HOSPITAL,

ON 3RD OCTOBER, 1887.

BY

CLINTON T. DENT, F.R.C.S.,

Assistant-Surgeon to the Hospital.

PUBLISHED BY REQUEST.

London :
HARRISON AND SONS, ST. MARTIN'S LANE,
Printers in Ordinary to Her Majesty.

1887.

HARRISON AND SONS,
PRINTERS IN ORDINARY TO HER MAJESTY,
ST. MARTIN'S LANE.

To

WILLIAM WADHAM, M.D.,

AND

TIMOTHY HOLMES, F.R.C.S.

THE NATURE AND SIGNIFICANCE OF PAIN.

To those who this day enter a school distinguished for something more than mere traditions. To those who renew their associations and mark off another stage in their life's work. To those (on whose good will and good estimation we so largely depend) who, having carried into practice lessons first learned here, uphold the honour and the credit of St. George's before the public. To those who by their presence show that the union of School and Hospital is not a mere wordy fiction; and lastly, to those who having ceased active work with us yet testify their undiminished interest in all that concerns our prosperity here. To one and all, in the name of the Lecturers, welcome! And indeed it is good to meet thus annually to take stock of our progress and our position, even though it result only in the infinitesimal good of making laudable resolutions. The happiest nation, proverbially, is that which has no history : surely then a medical school should be counted most fortunate that has no change to record. But since we last met, if there has been no actual, still there has occurred what may be termed a potential, change, inasmuch as two of the most important offices in the school are in new, though by no means untried, hands. What a school owes to its Dean you all know, but it is difficult for any of us adequately to realise our real obligation to our late Dean. It is beyond my power to express it in words. The expression must not be taken in a financial sense if I say that we owe also much, very much, to our late Treasurer, whose wide experience and judicial mind have exercised no less valuable influence on the conduct of the school, than on the surgical practice of this hospital and this country. For the moment I am more concerned with retrospect than with anticipation, and am glad to think that our senior colleagues are still associated with us, though the offices they have held so long, and whose duties they have discharged so honourably, have passed into other hands. Let us at least pay one tribute of gratitude that I venture to think will be appreciated, and strive stead fastly and with energy to carry on, through good or evil report, the work on the lines which our late Dean and Treasurer have laid down, and which have been so successfully pursued for many years. So may the invocation of Horace apply—

Di, probos mores docili juventæ,
Di, senectuti placidæ quietem
Date.*

On the subject of one, and that a most important addition that has been made during the past year, namely, the new physiological laboratories, I may not now speak, for that has already been dealt with by abler hands than mine in the eloquent address we listened to with so much pleasure from Mr. Pollock last April. The pupil fears to rush in where the master has dared to tread. I do not care to give you a chance of drawing an unfavourable comparison with those who have preceded me in the task devolving upon me to-day. Moreover, I have no good counsel ready to hand to give you; holding to some extent with the cynic who said that advice is almost the only commodity which the world is lavish in bestowing and scrupulous in receiving, although it

* *Carmen Sæculare.*

may be had gratis, with an allowance to those who take a quantity. In truth, gratis advice seems nowadays reserved for the public, and is unsuited to the profession. Nor will I invite your attention to any phases of the burning question of medical education and medical diplomas. These matters were involved when I first became practically interested in them : they are now chaotic. Let us assume for the moment that integrity and straightforwardness in the profession are of more moment than the details of the steps to be taken that will entitle a man to write any particular letters after his name. Honesty is a good policy, and should not be less revered because, to adopt the language of Insurance, it is a policy on which, in practice, a largish premium has to be paid.

These subjects, however attractive, I may not now touch. Abuses there are, and to spare. Genuine reform and true progress are still needed. But it is easier to criticise than to contribute, and more, vastly more, common. Let me for a few minutes endeavour to turn your thoughts to widely different matters.

I suppose that no one would question that, as the very *raison d'être* of medicine as an art must be reckoned the relief of Pain. It is but a paraphrase to say that the cardinal aim is the mitigation or cure of disease : for that is but dis-ease, want of ease, pain. The introduction of such topics needs no excuse, for day by day a Physiology of the mind assumes more practical shape. Between Physiology and Psychology there is still a great gulf : the one science begins long after the other has finished; but year by year, as physiological knowledge advances, the space is narrowed, and the time must arrive when the two sciences will begin to merge one into the other. Moreover, in our curriculum we are apt to lose sight too much of the ultimate objects of the Medical Art in the feverish craze for detail at the moment; and yet once in a while it is good to catch a glimpse of the goal towards which we are striving. I have known the sight of the summit of the mountain, albeit very far off, stimulate the wearied climber to fresh exertions.

The etymology of the word pain alone furnishes a striking instance of its universality. In a vast number of languages the word is identical, and, as Wedgwood remarks, the Latin word from which it was derived was no doubt enabled to spread itself so widely from the prominence of the idea of retribution and punishment in religious teaching. Nine-tenths of written history is but a record of the varied

sufferings of mankind. In the religious development of the world it was considered, it is
still considered by some, more efficacious to
presuppose the post-mortem continuance of a
nervous system, and to appeal to the dread of
pain, than to any other argument. In mediæval
ages, when men tortured and burnt the body,
pain was the common means adopted to arrive
at the truth ; I seek to turn the tables now and
invoke the scientific truth to explain the pain.
Whence comes this universal liability to pain in
mankind, and what is its value and significance ?

Without assuming for the moment any theory·
of evolution, we may turn profitably to the lower
animals for the first indication of an answer.
.For what is the apparatus required to constitute
a sentient being ? A nerve beginning to receive
the impression from without ; a nerve to conduct it ; and, most essential, a brain to realise it.
Probably in the class of which the sea anemone
is an example we may find the first indication of
a nerve, for we consider that to be a nerve which
has the functions of such. In this animal the
nerve, if such it be, is represented by a streak
of homogeneous matter, a process, as it is called,
jutting out from a cell placed close beneath
the integument. A stimulus applied to the
extremity of the process produces a change in

the central cell, and yet the process itself under-
goes no apparent alteration. Here we have a
nerve, but no brain, and therefore no sensation.
As we ascend the animal scale similar struc-
tures are found more and more distinctly. The
streak of homogeneous matter which is the
essential part acquires a covering sheath, and is
more easily detected; gradually it assumes the
form of the familiar white thread recognised as
a nerve. In animals in which the nerves are
more distinct the cells too are more abundant and
collected into masses; gradually some of the
masses become larger in proportion to the rest,
and occupy constantly the same position in the
head. Next they fuse together more or less, and
the resulting collection is called the Brain.
But the essential part of the nerve proper
remains still the same, similar in appearance
and identical in function; between the cell
process of an anemone and the essential part of
a nerve in man there is no difference. The
brain, in proportion to the rest of the nervous
system, becomes larger and larger, and more and
more complex, until finally in man it reaches its
highest development. At some point or another,
as we range over the animal scale, this collection
of nerve matter which we call the brain first
acquires the dawning power of appreciating the

impressions of which the nerve permits the con-
duction.   Not until that condition is fulfilled is
the possessor a sentient being.   At some higher
point which cannot be fixed with any precision,
the property first creeps in of realising the impressions; what was below this merely a sensation
henceforth becomes the idea of a sensation, at
first faint and scarcely recognisable, gradually
assuming clearer and more decided form.

Now pain is simply an intense or disorderly
sensation, or one that interferes with health or
comfort.   What degree of sensation constitutes
a pain cannot be laid down with precision, for
it is but a question of degree.   So it is with
individuals.   Most of us know some people
whom we consider commonplace ; others we
class as eccentric, others again as mad, but we
cannot define exactly the point where the commonplace individual becomes eccentric, nor again
the amount of eccentricity which entitles a
person to be considered mad.   There is a large
borderland between any such artificial varieties.
Our minds are absolutely incapable of appreciating impressions travelling to the brains of
others, and we but estimate character by our
own degree of intellect.   Thus to the commonplace person an eccentric man appears a highly
singular individual, but between a person with

enough of character to have natural oddities and the eccentric being there is a fellow feeling. Again, if we look at the colours of a spectrum, the difference between the red at one end and the blue at the other is obvious enough, and yet we cannot draw any sharp line where the red passes into orange or the blue shades off into violet; and so it is with sensation and pain. If we hold a moderately warm object in the hand the sensation of slight heat is pleasant; gradually raise the temperature of the object grasped, and gradually the sensation will change in character. "Above a certain point," says Sir W. Hamilton, "the stronger the sensation the weaker the perception." But before any decided stage of pain is reached there will be an interval during which the mind, which is located in the brain, is unable to determine whether the sensation be one of pleasure or pain. Here, already, we have sufficient proof that pain is but an exaggerated sensation, and that the privilege of pleasure is accompanied by the liability to pain. Nor can the mind, again, accurately determine the point at which health or comfort would be interfered with by the intensity of the sensation, or, in other words, the point at which pain begins. We are only enabled, in short, to quickly determine when the degree of sensation excited is decided and abrupt, and the standard will vary with different individuals. We all know the April condition of mind, half laughter, half tears : Robson the actor was famous for his power of keeping an audience in this state, a faculty that certain writers, notably Charles Dickens and Bret Harte, possess also in a marked degree. Here the brain is reached through the special senses of sight and hearing, and the soul of the brain, the mind, cannot tell whether it is pleased or pained. Impressions derived through sensory nerves which are not special may have the same effect, so that the brain cannot determine whether it is tickled or jarred. We experience a sharp thrill on hearing any sudden news ; analyse the sensation and it will be found to be identical whether the news be good or bad ; subsequent thought leads us to decide whether we call the thrill one of pain or one of joy. We acknowledge the same fact when we speak of particular sensations as being " acutely " pleasurable.

In the exercise of our profession, however, we are but little concerned with sensations that are on the borderland. There for the most part the impressions are unmistakable. Not that the mind is able to realise pain as such, but only,

as it were, to become aware that it is disturbed
by a series of impressions too disorderly or intense
for the comprehension. And in this, possibly,
we may catch a glimpse of the nature of pain,
as being the unavailing effort of the consciousness to realise its own disturbance. For we are
so constituted that it is actual pleasure to realise
a simple sensation : if the stimulus be a little
more complex or intense, so long as extra effort
is able to meet it, the conditions will not alter.
And here we perceive the source of the pleasure
derived from acquiring knowledge. But when
a point is reached at which the mind cannot
realise the sensations brought to it, there results
but disorder and confusion. And here we see
the source of the distress resulting from failure
to understand what is yet intelligible—a form of
anguish to which it must be confessed people
are able rapidly to inure themselves.

In medicine and surgery the vast majority
of pains are due to intense impulses transmitted
to the brain along sensory nerves, for the ordinary sensory nerves are the most numerous.
Obviously, however, sensory nerves wherever
situated may transmit impulses giving rise to
pain, or, in ordinary parlance, any part with
sensory nerves is liable to pain. So that not
only the portions of the body having nerves

concerned with the sense of touch, but others such as the great viscera—for example, the stomach and intestines—are not exempt. Of many parts of the body ordinarily we are but little conscious when they are doing their work naturally, as the heart, lungs, and liver, and the mere realising by the brain of impressions from these sources constitutes evidence of their disturbance. Consciousness of the beating of the heart implies some disorder of that part, functional or organic. The man who knows he has a liver will not unfrequently have enough mental disturbance to convert him from a human being into a misanthrope. Even so when we are aware that we have eyes or ears we have evidence that these organs are failing : here the consciousness is really that of the extra effort required in the endeavour to maintain a previous standard, for the person who has been myopic or astigmatic all his life is often unaware of the fact. And in this case there seems to be an exception to the rule that natural degeneration is not painful. The phenomena of living muscle afford an admirable instance of the graduation of impressions leading up to pain. Muscular contractions are perpetually going on in us. Of the multitudinous slight movements giving rise to the respiratory act we are unaware, through

familiarity, unless we give attention to their
occurrence. Rather more powerful contractions
voluntarily performed give pleasure—" Rejoiceth
as a strong man to run a race," says the Psalmist.
Hence the attraction of athletic exercises, ex-
pressed in physiological terms. Of its truth some
of you will probably find opportunities of giving
practical demonstrations. Putting all the force
possible into muscular contractions by great
effort of the will leads to sensory impressions
so vividly felt as to be not far short of painful.
The physiognomist would find that the expression in a man's face finishing a hard mile race
was similar to that of a person undergoing torture.
Finally, intensely powerful muscular contractions independent of the will or control send
up impulses which, being too acute to be realised
as definite impressions, become horribly painful.
In lockjaw or the convulsions of strychnine
poisoning the agony is hideous. So, too, where
the utmost efforts are made by the muscles to
ward off suffocation the suffering is propor-
tionately intense.

Pain is not merely confined to ordinary
sensation or to the special nerves, if there are
such, of the sense of touch. Of the remaining
senses with their special nerves the conditions
are the same. They discharge their functions

B

in the same way. The impulse in one case
giving rise to the sensation of touch, arouses in
the brain the idea of light in another, of sound
in the third, and so on. Excessive stimulation
will lead to pain. Thus it is painful to gaze at
the sun or at an arc electric light. A sound of
great intensity or shrillness, or the confusion of
'a musical discord, is absolutely painful. To the
sick person with brain disordered directly or
indirectly, slight sounds or degrees of light,
which would be disregarded in health, become
intolerable : for in illness one of the first
qualities lost is that of inhibition. Now,
through the special senses of sight and hearing
the great bulk of our education is conveyed to
the brain. It should and does follow that
education and intellectual development render
us susceptible to new pains derived for the most
part through these special senses. We admit
as much when we talk of painful sights. But
this epithet is often used loosely. A severe
operation might, for example, be considered a
painful sight. A person watching it for the
first time will often wear an expression identical
with that induced by pain derived from other
sources, as by irritation of a sensory nerve. He
may even faint away as he might faint from the
severe pain of a sprained ankle. The observa-

tion too is common that on those of coarser intellect sights such as a surgical operation have little or no effect. But that the term painful is not strictly accurate is shown by the ready manner in which the mind accustoms itself to such ordeals as witnessing an amputation, or the carnage of a battlefield. To the educated and sensitive eye of the artist, ill-assorted combinations of colours may cause actual pain, and so may inharmonious sounds to the educated and sensitive ear of the musician. To the trained eye of the anatomist or physiologist the sight of unnatural deformity, such as is often produced in obedience to the dictates of fashion or vanity, will give actual pain. To such impressions neither the artist, musician, or man of science will become accustomed or less sensitive. With the sense of touch the conditions often appear to be different when they are not so in reality. Thus, the blacksmith inures himself to handling hot substances by modification of the integument covering the nerve beginnings. Here, though the nerves may conduct as well, or the brain be disturbed by pain as efficiently, the impressions are partly stifled at the onset. The person who is highly sensitive to pain is, as a rule, in fault, not in the nerve paths, but in the brain. Alteration, diminution of resistance

in the nerve path is the least common cause of
the condition.   Alteration in the nerve com-
mencements implies little more than that they
are unduly exposed or modified by disease or
injury.   It is in the mind that the impressions
are commonly misinterpreted or exaggerated,
and naturally, for it is in the mind that the
delicate task of converting them into ideas is
performed.   And so ofttimes when we talk of a
tender finger we ought to speak of a tender
brain.   Some emotional people whose nerve
path may be healthy enough, and whose trouble
is functional and solely in their brains, are wont
to speak of themselves, when fishing for
sympathy, as "masses of nerves."   They have
no more nerves than any one else.   They
ignore the part really at fault, namely, the
brain.   Perhaps, in a sense, they are right, but
the expression sounds strangely ironical to the
physiologist.

It is in the brain, then, that we must chiefly
look for the explanation of the varying degrees
of sensibility, and as there are all grades from
the highly intellectual to the very stupid so are
there all shades of sensitiveness from the highly
emotional to the excessively callous.   Education
will only mend the condition within certain
limits ; it can only brighten that which is capable

of taking a polish. Many minds can only be dipped into education, and such electro-plated intellects won't stand much friction. Still, if we can educate ordinary sensation or touch we can educate the sense of pain. This is precisely the course that the so-called hysterical people go through. With their ideas all turned in on themselves and not projected outwards they do cultivate and educate their sense of pain. The pain will be as real in such persons as in others; for subjective may at least be as vivid as objective sensations. Hysterical, neurotic folk have their own views of what is logical and ask for no proofs; what they feel, they assume exists. The saying of Aristotle, "*Nihil est in intellectu quod non prius erat in sensu*," is only nearly true : it does not hold good of such disordered intellects. Neither is it absolutely true, though more nearly correct, to say, *Nihil est in sensu nisi est in intellectu.* Take the instance of a person who after a head injury hears constantly singing noises. There are no objective vibrations to cause the sensation : yet they are no less real to the patient than they would be if caused by a kettle "singing" on the hob. There is no need to multiply instances, for hallucinations of the special senses are of the commonest occurrence, and may be of all degrees. So that if

pain is but an intense sensation, it follows that
pain too may be subjective or objective. And,
reasoning on another line to the same point, we
may see that such is the case. Some action,
some disturbance we suppose takes place in the
brain as the result of an impression actually
brought to it from without. If a precisely similar
action or disturbance originated within the brain,
clearly the same sensation would be realised; and
if the action were excessive, the same pain. A
brain may be functionally as unduly tender as
any given part of the body. We recognise
this in fatigue or illness; we do not often
enough recognise it in what is called hysteria.
The world is sorry for the person whose brain
is disturbed by the impressions coming from a
disorganised joint; and, too often, only laughs
at the patient who has the same disorder
brought about by a different condition. Yet
both are diseased. The pain is but objective
in the one, and subjective in the other. Stranger
still, we sympathise with the lunatic who believes
that he has committed the unpardonable sin,
or who supposes that every one is conspiring
to poison him, while we are apt to make light
of the suffering a patient actually endures who
imagines she has an acute disease of the spine;
and yet indifference or mockery may be to the

tender brain as a blow or jar to the tender joint. If I may be permitted to digress for a moment in order to draw a practical conclusion from these remarks in the shape of advice, it would be to urge the study of the mental aspect of disease. The subject is included already in the vast curriculum that lies spread before you, and so I do not advocate the introduction of any new item. That may be safely left to Boards of Medical Education. The temptation to add new subjects is only, it would seem, too attractive. For medicine may lay under contribution with advantage almost every science—save, perhaps, Pure Mathematics and Botany. But for those of you who have the immense advantage of a little extra time in your student career beyond the narrow limit prescribed, the study of insanity in asylums will prove of a benefit in ordinary practice that you can hardly over-estimate ; and I would gladly welcome any arrangement that might be made whereby students in general hospitals might have facilities for working for a while in asylums : asylums they are no longer, but hospitals in the best sense of the word, with the unique merit of being special hospitals against which nothing can be said.

Hitherto I have but dwelt on the darker

side of the picture, and have spoken only concerning the sources and significance of pain. It is a relief to turn to another aspect of the subject and note the conditions which diminish suffering. Pain is confined to the animal world, the liability to it being proportionate to the anatomical and, still more, to the physiological development. As therefore the liability increases, so does the capacity of the intelligence to minimise pain augment. Pain, as a symptom, bears no fixed proportion to severity or importance of disease. A person, for instance, may suffer torments for years from neuralgia, a condition in which after a while the apprehension scarcely renders the patient less miserable during the respites from actual suffering. And all this neuralgic pain may be due to a little disease of one tooth or a minute outgrowth of bone pressing on a branch of a nerve. There seems something wasteful, as it were, in such cases, and we not unnaturally consider it almost an evidence of the cruelty of Nature. What we don't understand, however, we are all apt to depreciate. On the other hand, though a cause almost trifling may evoke a mighty amount of pain yet scarcely affect the duration of life, many grave diseases will run their course free from it. A person may have a great aneurism of a most important blood vessel and know nothing of it. The liver may be riddled with abscesses, or saturated with cancer, and yet give rise to no pain. The lungs may be so diseased in advanced phthisis that there is scarcely any sound tissue to be found. Yet in the consumptive there may have been, practically, no pain ; often only an increasing brightness and hopefulness of mind as the hopelessness of the physical condition advances. There is a poetry in such natures of which Death is but the final stanza ; gently claiming his own while the bodily mind is still busy with projects and plans for this world. So here, if Nature takes, Nature also gives.

It is a noteworthy consideration that the value and use of pain is of importance chiefly at the moment. For we have no real power of recollecting pain. When passed, no impression is found to have been made on the memory. By no effort of the mind can we reproduce at will a condition of pain. With ease can we recall in memory sights and sounds so vividly as to deceive ourselves with an appearance of actuality. While then we have, as it were, a mind's eye and a mind's ear, we have no equivalent power as regards common sensation.

Most of us have no real "touch memory."* Such feelings cannot be revived spontaneously, and the nearest approach that we can make is to abandon our minds to disquieting thoughts. Nor is this surprising; for we can but recall in memory that which has been sharply and clearly impressed on the mind, and pain in its essence is disorderly and confused, so that we can no more revive it than we can feel cold by thinking of cold. We can imitate the action of shivering which actual cold would induce, as we can reproduce any other expression of pain in more or less recognisable form; but this is only mimicry, and the portrayal of an effect does not amount to a revival of the condition.

The observation has been made—and it is a remarkable one—that whereas pain is confined to the animal world, so the remedies against it are almost wholly drawn from the vegetable kingdom. But we shall be taking far too narrow a view if we seek only in the discovery of fresh agents and drugs for evidence of the progress of medicine in its

---

* The expression is not a good one : but I can think of no better, and the context, I hope, makes my meaning clear. The quality, I suppose, only lies dormant. In the blind, apparently, it is developed to some extent.

capacity for relieving or curing pain : for such remedies aim only at the mitigation of a symptom. While medical art can, it is true, often do no more at present, it is ever striving, and striving successfully, for something higher. Medicine, as distinguished from surgery—an artificial separation to which you will find less attention paid when you quit the hospital work —makes its progress so much in the direction of prevention that its advance is less obvious to the unheeding. Even thus men are ever more keen to hail some individual as a brilliant discoverer than to acknowledge the patient toil of those whose efforts of mind he has but focussed. Yet the value of science is not to be measured by the fame that it brings to one, but by the benefits it confers on all.

Surgery, however, has to do with another aspect of pain ; in that, in the shape of surgical operations and procedures, a vast amount of pain is deliberately inflicted in order to bring about by calculated means ultimate relief. It is difficult to conceive the enormous amount of pain that must have been endured. For more than two thousand years at least in the world's history have we authentic records of the performance of surgical operations. Of lithotomy, of amputations, of plastic operations, and many

others of the first importance, we have indubi-
table evidence. Moreover the suffering entailed
was added to by less accurate knowledge of
physiological laws, and augmented often by
superstition, as for example by the Arabians
in whom the dread of loss of blood was so
great as to lead to the use of the actual
cautery in operations to an extent that can
hardly now be realised.   Men recognised the
terrors and endeavoured vaguely to obviate the
pain—for they felt the pain was preventable—
but with little success.   Opium and other
drowsy syrups were tried, and found wanting.
The Oracle of Science was invoked, but gave
no answer that men would read.   Yet nearly
ninety years ago, one Humphry Davy, then a
youth of twenty-two, in an essay concerning
Nitrous Oxide Gas, wrote—"As it [Nitrous
Oxide] in its extensive operation appears capable
of destroying physical pain, it may probably be
used with advantage during surgical operations
in which no great effusion of blood takes
place."*   But the hint passed unnoticed.   It
was not till many years later that the value of
ether and chloroform as anæsthetics was first

* "Researches, Chemical and Philosophical, chiefly concerning Nitrous Oxide," by Humphry Davy (London, 1800)
p. 556.

made known. Into the details of the history of
the discovery I may not enter, for the story has
already been admirably told by Sir James Paget.
The introduction of ether has made, or should
have made, the names of Morton and Warren
household words, though those of Long, Wells,
and Jackson are now almost forgotten. It
marked an epoch. For thousands of years the
grim spectre of Pain had stood an unfailing
attendant over all surgical proceedings, and now
it was laid, and for ever. So universal is now
the use of anæsthetics, so much is their employ-
ment a matter of almost routine, that it appears
incredible that they have only been vouchsafed
as a gift of God to man for the last forty years.
Yet it was only on the 16th October, 1846,
that Morton gave the ether and Warren performed the first surgical operation of any
magnitude under its influence.* The news
of the discovery spread with amazing rapidity.
In London, ether was probably first used at
University College Hospital, by Mr. Liston,
on December 21st, 1846,† and on January

* *Lancet*, 1847, Vol. I, p. 6. See also Sir James Paget's
essay, "The History of a Discovery," in the *Nineteenth
Century* Magazine for 1879.

† *Lancet*, 1847, Vol. I, p. 8. Mr. Robinson was actually
the first to give ether in London on 19th December, 1846,

14th, 1847, Mr. Henry James Johnson* in this Hospital amputated the leg of a man placed under its influence.† So strange and novel was this new departure, and so much interest did it excite, that on the occasion of the operation the theatre could not contain the spectators desirous of witnessing it, so that many crowded on to the roof, and saw what they could through the skylight.

No greater discovery has ever been made for the benefit of mankind, but from the fact that it prevents pain its real importance is apt to be underrated. Those who, like myself, have witnessed but few surgical operations performed without the use of anæsthetics cannot readily, or without thought, appreciate what was the condition of things before their introduction: how the apprehension must have added to the terror, or how the mental effort to endure must have exhausted: how in many the courage must have failed at the thought of the price in pain to be paid for relief, and how therefore

for the extraction of a tooth. *Cf.* Snow, "On Anæsthetics" (London, 1858), p. 18.

* Afterwards known as Mr. James Johnstone.

† *Lancet*, 1847, Vol. I, p. 104. Possibly ether was not used for the first time at St. George's in this case. See Dr. Merriman's remarks, *op. cit.*, pp. 99, 100.

diseases were allowed to drift on which might have been otherwise alleviated or recovered from. Anæsthetics are sometimes spoken of as if given only for the convenience of the surgeon, and thus again their real significance overlooked, even as the therapeutic value of a bed as a medical appliance is ignored. Enormous, beyond all reckoning indeed, is the boon of the use of anæsthetics in surgery to mankind. Yet it must not be forgotten that these agents have their risks ; for if the danger be borne in mind these risks will be diminished. They are given a thousand times with success, but once in a way, perhaps directly and with awful suddenness, perhaps indirectly, as by the induction of bronchitis, deaths do occur. It is mere fatalism to speak of such calamities as inevitable. We may yet reach nearer to perfection if we do not assume that we are already there. Indeed, that there is room for further development in this branch—*i.e.*, the prevention of pain inflicted with an object—the recent introduction of cocaine shows. The history of this drug is instructive, for it shows how by a physiological misinterpretation the real value was overlooked. Used by the mountaineers of the Andes to enable them to take long journeys without food, the drug was supposed by those who first studied it—among

whom Sir Robert Christison* may be mentioned
—to have special life-supporting powers. In
this respect actual experiment proved it to be
found wanting. As a matter of fact coca but
obviated the pangs of hunger, much as the common traveller's device of tightly girding the
belly impedes the upward flow of the disordered
sensory impressions to the brain. In like
manner we compress, mechanically, a part that
has just been injured. Almost accidentally,
and only three years ago, Kohler discovered
the real anæsthetic properties of the drug, and
its use is now universal. In this hospital it
was first used for an eye operation by Mr.
Carter, on 16th November, 1884. Everything
has been said we hear, everything has been
tried; but rest assured that

> " Words, like Nature, half reveal
> And half conceal the soul within "

and the full significance of many experiments
still lies buried, waiting to be unearthed and
set forth in the light. Seek, then, and you
shall find.

---

* Yet Christison wrote down, " Experimental enquiries
. . . prove that in small animals cocaine produces in
an adequate dose *paralysis of sensation*, tetanic convulsions
and death."—*British Medical Journal*, 1876, Vol. I, p. 529.

Much has been said, and books, or, at least, a book has been written, on the "Mystery of Pain." There is no harm in the term if it be used in the old sense, and not simply as a barrier to turn back the scientific inquirer. To hold that it is such in its nature is neither logical nor physiological. Is not the "mystery" merely a problem difficult of solution, and then only by slow and laborious means, not an enigma which is either to be guessed or given up? If only we may learn something of the laws which govern pain, something of the manner of its working, we are yet doing much. True that pain is often a misleading guide to those who do not appreciate its physiological significance. Our reasoning may frequently be at fault, for pain as a symptom often speaks in language which is obscure and difficult to interpret. But in such instances with whom is the fault? When the intellects are but limited in all, it is merely a truism to say that he succeeds best who makes the fewest mistakes. Though we may fail from imperfect deduction in any or in many given instances, yet still, if we work systematically to the advancement of knowledge we may employ the same advantageously in other and simpler cases. Meanwhile the person who blunders along at haphazard ignores

c

all failures till he stumbles perchance on some
dazzling theory which has the effect of mentally
blinding him.   In gold-digger's phrase, then,
wash every scrap of your information with
care, and labour for the grains of ore : for truth
is not found in nuggets, and there is no room
for conceit in science.

While we rightly seek, therefore, in our art
to diminish by all means the amount of pain
that has to be endured : while we, as education
and civilization advance, strive with all our
power to palliate an inevitable consequence of
progress, and while we search for new or more
perfect methods of attaining such ends, let us
not yet forget that pain has its own uses and
value.   It is an old factor in the development of
mankind, as those may learn who will but peruse
the third chapter of Genesis.   The existence
of pain is of the highest defensive value to the
individual; further, it prevents, as has been
said, the undue use, that is the abuse, of our
physical attributes.   Should we willingly part
with any of our senses because through their
agency impressions may be derived which are
unpleasant ?   Shall we desire not to see because
our eyes may rest on much that is hideous, or
wish to lose the sense of smell because in this
world the sweet savours are out-numbered by

the stinks? As well might we desire to have no power of motion because the muscle which can provide this for us may be afflicted with cramp, or become paupers in order that we may be exempt from taxation. Such may be the views of the Pessimist school, moral squinters who divide their time between complaining that there is no good here below and denying that there is any hereafter, but should not be of those who are possessed of healthy minds. Sensibility is our greatest privilege. Are we to complain because our senses are not rigidly limited and eclectic? Let us develop and improve or die. The more we educate our special senses, the more pleasure shall we obtain through their agency, and the more liability to pain. It is well for us if we do but bear in mind the profit gained. It is well for us, and it is well for our fellow-creatures; for as with a wider range of sensations comes a positive gain of happiness for ourselves, so may we utilise the disadvantages learnt for the benefit of others. For so shall we enlarge our sympathy, a quality which all mankind most earnestly desires. Conceive it possible that sympathy were non-existent, and the pains and sufferings of humanity would be trebled and quadrupled till they were past all bearing. For men and women will willingly submit to discomfort amounting almost to pain in their thirsty craving for sympathy. The word is truly derived and pregnant of meaning, for if we are concerned for another's pain we do actually suffer with him. So that sympathy may be as a bridge that marks a division between and yet unites subjective and objective pain. The Medical Profession is supposed by the unthinking—popularly supposed, in other words —to be insensitive to pain in others, especially the surgical branch of the Profession : that is as much as to say they have no feelings of sympathy. Familiarity is supposed to have begotten its proverbial offspring. People have unwittingly adopted for the type the characteristics which Celsus recommends when he says, " The surgeon should be young, or at least not advanced in years, . . . bold, unmerciful, so that, as he wishes to cure his patient, he may not be moved by his cries to hasten too much, or to cut less than is necessary. In the same way let him do everything as if he were not affected by the cries of the patient." Anæsthetics have changed all that, and nothing now could be further from the real state of the case. Only the very unwise could entertain a contempt for that which they but imperfectly understand, and in which we all are hourly endeavouring to advance and render more precise our knowledge. We but make scientific use of pain as a symptom wherever and in whatever form it is met with, and the more exact the interpretation of the phenomenon the more efficacious will be the relief afforded. But sentiment must follow after; for it will do less for the sufferer than skilful medical aid. It is not the moment to condole when the femoral artery is divided and spirting.

So we may recognise that pain is of high value and advantage from something more than the physical point of view; but this is not the time nor place to enter into the moral aspect of the question.

There is a tendency among the young—it is not difficult to see how it arises, especially to those who have studied in German-speaking schools—to assume that all the patients you meet with in the wards and the out-patient rooms are there in virtue of some law of Nature : that they are material to supply the the demand for instruction : that people fall ill as leaves wither. A moment's reflection convinces you that such views are not justifiable. We hear much, and rightly, now-a-days of the grievous waste of life from preventable diseases.

Disease and pain are for my present purpose
convertible terms, and in the clinical work
of a large general hospital we should regret
no less the waste of energy, the squandering
of the utility of existence, the failure to profit
by the opportunity of life, which is incessantly
before our eyes.   The art of medicine does
not consist in mere germ-hunting.   We need
not in our daily work seek further back for
the origin of evil than in Poverty, Ignorance,
and Vice.   Formidable antagonists, in truth :
but though vast and terrible, these monsters
are not unconquerable.   The sum of pain
that has to be endured by mankind is huge,
and we still grapple with it but inefficiently.
Yet, as Education progresses, so must one,
if not two, of these factors in causing disease
and suffering—Ignorance and Vice—shrink
back before it.   For Ignorance and Innocence
are blood relations ; but Ignorance and Vice
are but associates always banded together
for evil.   With Poverty as a source of pain
you will have only too frequent occasion to become familiar—so commonly, in fact, that you
may be apt to neglect the factor as a cause.
Go into the wards, look into the out-patient
rooms, and in one form or another it meets your
gaze everywhere.   Here is a poor woman with

a painful ulcer of the leg, the result mainly of want. Some slight injury has been perforce neglected, or treated by cold water or poultices, and repair cannot take place in the enfeebled tissues. " Rest the leg," she is told, " and the pain will subside and the ulcer heal." " Rest! That means loss of work and loss of money ; and I have a husband and "—fill in the number at discretion—" children." Then comes an appeal for strengthening medicine. Finally, the patient gets a bandage, perhaps accompanied by the grimly ironical counsel to live well. Or she gets her medicine, a " tonic" it may be, bitter stuff to create an appetite which she has no means of allaying. So she goes on, and the pain goes on too, and the world turns round as before. And with such you will see a crowd of ricketty children, cases of joint and spinal disease, incipient phthisis, and what not. Poverty and pain. You hear often the fallacy urged that health cannot be bought. But much immunity from pain or disease can be bought, for rest is the natural antidote to pain. Where would be such cases as I have described, and where the thousand and one deformities met with in our everyday work here, if rest could be provided, not merely prescribed ? Such patients could be taken in and put on the road to recovery, or at the outset cured, at a shilling a day, if there were adequate hospital accommodation and means to buy health for the poor. Give us the shillings. So might be alleviated or removed a vast amount of preventable pain. , There is a huge dead weight of it in this world. Let each one who has the proper restlessness of health and the energy of a sympathetic mind shove against it, and it will yield, if ever so little. There is ample room for more workers, and so I rejoice to meet the band of recruits who enlist to-day in this work.

My address is done. We have been sailing hitherto in waters in which currents and counter-currents of opinion have often raised the sea of controversy; and with relief may I now glide into the smoother channels of convention-ality. As it was my pleasure at the outset to welcome you at your coming, so is it now my privilege as we part, each to take up his share of the work of a new session, to give you God-speed on your journey. I trust that you will not hold me guilty of striking a false note at this, the very outset. To steer between a didactic lecture on the one hand, and mere philosophical babble on the other, is no easy task. I have led you deliberately into some-what speculative fields of thought, believing

that it is good once in a while, on occasions
such as the present, to exercise the reasoning
powers and test their suppleness and elasticity.
For we should not be content to use the
mind as a mere lodging-house for facts, to
accommodate—on a short tenancy—all that
can be crowded into it. The man who uses
his intellects but in such a direction commits
the physiological mistake of directing all his
energies to absorption. Let him from time
to time also secrete a little, and the balance
will be better kept.

You enter what is often spoken of as a
toilsome, laborious profession. Thank Heaven
for it. There should be no torture to the man
of healthy mind greater than to be bound hard
and fast to a life of ease and sloth, with the
ceaseless drip of ennui falling ever on his
rusting intellect. You will hear perpetually
that the Profession is inadequately recognised
in the matter of worldly honours. Judge
for yourselves how far this is matter for
regret. But I would not have you without
ambition. Your profession should be part of,
as it were a graft on, your life, not a mere
staff to support it ; and life without ambition
is merely existence. Do you hold that worldly
honours seem partially bestowed ? What of

D

it? If you think but of them you will only
pursue a dangerous road that leads towards a
goal of disappointment. The Medical Profession demands of its members no more, but no
less so-called drudgery than other lines of life.
You will hear much of what is spoken of as
carpet practice; but most of it is on the bare
boards. The world may resound with the fame
of those who have known how to "touch the
magic string," while it may seem to leave out
in the cold and ignore those noble members of
our Profession who, toiling at laborious parish
work, or in ill-paid practices in mining or poor
country districts, are striving their utmost to
combat disease and pain. But are such
without recompense? Said a great writer, one
who in herself to the vigour and breadth of
a man's mind added the quicker and more
sympathetic observation of a woman—" The
growing good of the world is partly dependent
on unhistoric acts : and that things are not so
ill with you and me as they might have been,
is half owing to the number who lived faith-
fully a hidden life, and rest in unvisited tombs."*
Words of truth, these. But even though the
world may seem to forget, let not the Profession
fail to acknowledge.

* George Eliot, " Middlemarch," concluding lines.

In whatever line fortune may hereafter direct you, may you so work as to deserve success : may the knowledge you gain here be used for the good of others, as it assuredly will be for your own : and may the training of this School be the one you can look back to with most satisfaction and gratitude, as having enabled you to win something better than mere success—the respect of others : and it is better to be respected of one than to be popular among ten thousand. So may your life be as epitomised in the old Persian stanza—

" On parent's knees, a naked new-born child,
Weeping thou sat'st, while all around thee smiled—
So live, that sinking in thy last long sleep
Calm thou may'st smile, when all around thee weep."

HARRISON AND SONS,
PRINTERS IN ORDINARY TO HER MAJESTY,
ST. MARTIN'S LANE.